Student Name:

Pre-Test Score:

Date Started:

Date Completed:

Final Test Score:

ADDITIONS

SAREEDO PRE-UNIVERSITY

Math Workbooks

Pre-Test:

1)
```
    3 1 2          2 4 5          1 7 5          4 6 9
  +   9 0        +   5 7        +   8 8        +   5 7
```

2)
```
    1 5 2          5 2 4          6 1 7          5 4 9
  + 9 6 7        + 2 5 8        + 6 8 6        + 7 9 8
```

3)
```
    7 1 2          2 7 4          1 7 6          2 8 5
  + 5 6 5        + 2 5 6        + 6 8 8          8 2 3
                                              + 6 5 9
```

4)
```
    6 1 9          2 1 4          6 7 4
  + 5 0 6        + 5 4 6        + 6 7 3
```

5)
```
    5 7 1 2        2 8 6 4
  + 5 6 5 9      + 2 5 6 7
```

Welcome To Your Class:

Keep Calm and  your Additions!

When you finish this class you are

an Addition Expert!

1) Add to 0, 1, and 2

(1) $1 + 0 =$

(2) $2 + 1 =$

(3) $3 + 1 =$

(4) $5 + 1 =$

(5) $4 + 1 =$

(6) $7 + 1 =$

(7) $9 + 1 =$

(8) $8 + 1 =$

(9) $6 + 1 =$

(10) $8 + 0 =$

(11) $9 + 1 =$

(12) $1 + 1 =$

(13) $7 + 1 =$

(14) $2 + 2 =$

(15) $1 + 2 =$

(16) $5 + 2 =$

(17) $7 + 2 =$

(18) $4 + 2 =$

(19) $6 + 2 =$

(20) $8 + 2 =$

(21) $3 + 2 =$

(22) $9 + 2 =$

(23) $7 + 1 =$

(24) $6 + 0 =$

(25) $7 + 2 =$

2) Add to 3, 4

(1) $1 + 3 =$

(2) $2 + 3 =$

(3) $3 + 3 =$

(4) $4 + 3 =$

(5) $5 + 3 =$

(6) $7 + 3 =$

(7) $9 + 3 =$

(8) $8 + 3 =$

(9) $6 + 3 =$

(10) $8 + 2 =$

(11) $9 + 3 =$

(12) $6 + 2 =$

(13) $5 + 2 =$

(14) $2 + 4 =$

(15) $3 + 4 =$

(16) $5 + 4 =$

(17) $7 + 4 =$

(18) $4 + 4 =$

(19) $6 + 4 =$

(20) $8 + 4 =$

(21) $1 + 4 =$

(22) $9 + 4 =$

(23) $7 + 2 =$

(24) $6 + 2 =$

(25) $8 + 2 =$

3) Add to 5, 6

(1) $1 + 5 =$

(2) $2 + 5 =$

(3) $3 + 5 =$

(4) $4 + 5 =$

(5) $5 + 5 =$

(6) $7 + 5 =$

(7) $9 + 5 =$

(8) $8 + 5 =$

(9) $6 + 5 =$

(10) $8 + 5 =$

(11) $9 + 5 =$

(12) $6 + 5 =$

(13) $5 + 5 =$

(14) $2 + 6 =$

(15) $3 + 6 =$

(16) $5 + 6 =$

(17) $7 + 6 =$

(18) $4 + 6 =$

(19) $6 + 6 =$

(20) $8 + 6 =$

(21) $1 + 6 =$

(22) $9 + 6 =$

(23) $7 + 6 =$

(24) $6 + 6 =$

(25) $8 + 6 =$

4) Add to 7, 8

(1) $1 + 7 =$

(2) $2 + 7 =$

(3) $3 + 7 =$

(4) $4 + 7 =$

(5) $5 + 7 =$

(6) $7 + 7 =$

(7) $9 + 7 =$

(8) $8 + 7 =$

(9) $6 + 7 =$

(10) $8 + 7 =$

(11) $9 + 7 =$

(12) $6 + 7 =$

(13) $5 + 7 =$

(14) $2 + 8 =$

(15) $3 + 8 =$

(16) $5 + 8 =$

(17) $7 + 8 =$

(18) $4 + 8 =$

(19) $6 + 8 =$

(20) $8 + 8 =$

(21) $1 + 8 =$

(22) $9 + 8 =$

(23) $7 + 8 =$

(24) $6 + 8 =$

(25) $8 + 8 =$

5) Add to 8, 9

(1) $1 + 9 =$

(2) $2 + 9 =$

(3) $3 + 9 =$

(4) $4 + 9 =$

(5) $5 + 9 =$

(6) $7 + 9 =$

(7) $9 + 9 =$

(8) $8 + 9 =$

(9) $6 + 9 =$

(10) $8 + 9 =$

(11) $9 + 9 =$

(12) $6 + 9 =$

(13) $5 + 9 =$

(14) $6 + 9 =$

(15) $5 + 8 =$

(16) $5 + 8 =$

(17) $7 + 9 =$

(18) $4 + 9 =$

(19) $8 + 8 =$

(20) $8 + 9 =$

(21) $3 + 8 =$

(22) $9 + 8 =$

(23) $7 + 9 =$

(24) $6 + 8 =$

(25) $5 + 8 =$

6) Add to 0-9

(1) $1 + 9 =$

(2) $2 + 5 =$

(3) $3 + 6 =$

(4) $4 + 7 =$

(5) $5 + 8 =$

(6) $7 + 5 =$

(7) $9 + 4 =$

(8) $8 + 3 =$

(9) $6 + 2 =$

(10) $8 + 1 =$

(11) $9 + 6 =$

(12) $6 + 4 =$

(13) $5 + 5 =$

(14) $6 + 7 =$

(15) $5 + 4 =$

(16) $5 + 3 =$

(17) $7 + 7 =$

(18) $4 + 6 =$

(19) $8 + 4 =$

(20) $8 + 5 =$

(21) $3 + 9 =$

(22) $9 + 6 =$

(23) $7 + 6 =$

(24) $6 + 7 =$

(25) $5 + 7 =$

0 + 1 =	0 + 2 =	0 + 3 =
1 + 0 =	1 + 1 =	1 + 2 =

1 + 3 =	1 + 4 =	1 + 5 =	1 + 6 =
1 + 8 =	1 + 7 =	1 + 9 =	1 + 6 =
2 + 0 =	2 + 1 =	2 + 2 =	2 + 3 =
2 + 4 =	2 + 5 =	2 + 6 =	2 + 7 =
2 + 8 =	2 + 9 =	2 + 7 =	2 + 5 =
3 + 0 =	3 + 1 =	3 + 2 =	3 + 3 =
3 + 4 =	3 + 5 =	3 + 6 =	3 + 7 =
3 + 6 =	3 + 3 =	3 + 1 =	3 + 5 =
4 + 0 =	4 + 1 =	4 + 2 =	4 + 3 =
4 + 4 =	4 + 5 =	5 + 2 =	0 + 0 =
4 + 8 =	4 + 5 =	4 + 3 =	1 + 5 =

1 + 1 =   3 + 2 =   4 + 3 =

2 + 5 =   2 + 7 =   2 + 2 =

3 + 3 =   4 + 4 =   5 + 5 =   3 + 6 =

2 + 6 =   0 + 7 =   3 + 4 =   1 + 5 =

2 + 7 =   2 + 6 =   4 + 2 =   4 + 5 =

4 + 4 =   5 + 4 =   2 + 6 =   3 + 5 =

2 + 4 =   2 + 5 =   5 + 1 =   5 + 2 =

3 + 3 =   3 + 5 =   4 + 3 =   5 + 4 =

3 + 4 =   3 + 5 =   3 + 6 =   2 + 7 =

3 + 8 =   2 + 6 =   3 + 1 =   4 + 5 =

1 + 1 =   4 + 4 =   4 + 2 =   4 + 5 =

5 + 4 =   5 + 2 =   5 + 3 =   3 + 0 =

4 + 3 =   5 + 3 =   5 + 2 =   1 + 8 =

6 + 1 =   7 + 2 =   5 + 3 =

8 + 0 =   8 + 1 =   7 + 2 =

5 + 3 =   5 + 4 =   1 + 5 =   1 + 6 =

1 + 8 =   1 + 7 =   1 + 9 =   1 + 6 =

2 + 5 =   8 + 1 =   9 + 0 =   2 + 3 =

2 + 4 =   2 + 5 =   2 + 6 =   2 + 7 =

2 + 8 =   2 + 9 =   2 + 5 =   2 + 5 =

3+ 0 =   7 + 1 =   3 + 2 =   3 + 3 =

3 + 4 =   3 + 5 =   6 + 2 =   3 + 7 =

3 + 6 =   3 + 3 =   7 + 2=   3 + 5 =

4 + 0 =   4 + 1 =   4 + 2   =   4 + 3 =

4 + 5 =   3 + 5 =   2 + 6   =   6 + 0 =

4 + 8 =   4 + 5 =   4 + 3 =   6 + 2 =

$5 + 6 =$ $8 + 2 =$ $4 + 6 =$

$6 + 6 =$ $7 + 6 =$ $7 + 2 =$

$8 + 3 =$ $7 + 4 =$ $6 + 5 =$ $7 + 3 =$

$4 + 4 =$ $5 + 5 =$ $6 + 6 =$ $7 + 7 =$

$8 + 8 =$ $9 + 9 =$ $6 + 7 =$ $5 + 7 =$

$8 + 4 =$ $7 + 5 =$ $9 + 6 =$ $2 + 7 =$

$2 + 8 =$ $5 + 9 =$ $6 + 7 =$ $8 + 5 =$

$3 + 9 =$ $3 + 7 =$ $3 + 6 =$ $3 + 3 =$

$3 + 4 =$ $6 + 5 =$ $4 + 6 =$ $3 + 7 =$

$3 + 6 =$ $8 + 3 =$ $3 + 1 =$ $9 + 5 =$

$9 + 1 =$ $8 + 2 =$ $7 + 3 =$ $6 + 4 =$

$5 + 5 =$ $4 + 7 =$ $5 + 6 =$ $1 + 9 =$

$8 + 8 =$ $8 + 5 =$ $4 + 3 =$ $2 + 8 =$

8 + 8 =	7 + 7 =	8 + 9 =
9 + 6 =	7 + 6 =	7 + 8 =

8 + 5 =	7 + 4 =	6 + 5 =	7 + 5 =
4 + 8 =	8 + 6 =	6 + 9 =	6 + 7 =
8 + 8 =	9 + 9 =	6 + 8 =	5 + 7 =
8 + 4 =	7 + 5 =	9 + 6 =	2 + 7 =
2 + 8 =	5 + 9 =	6 + 6 =	8 + 5 =
3 + 9 =	4 + 7 =	5 + 9 =	5 + 6 =
5 + 5 =	6 + 6 =	4 + 6 =	3 + 7 =
3 + 6 =	8 + 3 =	5 + 5 =	9 + 5 =
9 + 1 =	8 + 2 =	7 + 3 =	6 + 4 =
5 + 5 =	7 + 7 =	5 + 6 =	7 + 9 =
8 + 8 =	8 + 5 =	4 + 8 =	8 + 6 =

7a) Add to 10, 11

(1) $9 + 1 =$

(2) $10 + 1 =$

(3) $10 + 2 =$

(4) $10 + 3 =$

(5) $10 + 5 =$

(6) $10 + 4 =$

(7) $10 + 6 =$

(8) $10 + 9 =$

(9) $10 + 7 =$

(10) $10 + 8 =$

(11) $10 + 9 =$

(12) $10 + 10 =$

(13) $10 + 9 =$

(14) $11 + 1 =$

(15) $11 + 2 =$

(16) $11 + 4 =$

(17) $11 + 5 =$

(18) $11 + 7 =$

(19) $11 + 3 =$

(20) $11 + 5 =$

(21) $11 + 8 =$

(22) $11 + 6 =$

(23) $11 + 9 =$

(24) $11 + 7 =$

(25) $11 + 10 =$

7b) Add to 12, 13

(1) $11 + 1 =$

(2) $12 + 1 =$

(3) $12 + 2 =$

(4) $12 + 3 =$

(5) $12 + 5 =$

(6) $12 + 4 =$

(7) $12 + 6 =$

(8) $12 + 9 =$

(9) $12 + 7 =$

(10) $12 + 8 =$

(11) $12 + 2 =$

(12) $12 + 10 =$

(13) $12 + 9 =$

(14) $13 + 1 =$

(15) $13 + 2 =$

(16) $13 + 4 =$

(17) $13 + 5 =$

(18) $13 + 7 =$

(19) $13 + 3 =$

(20) $13 + 5 =$

(21) $13 + 8 =$

(22) $13 + 6 =$

(23) $13 + 9 =$

(24) $13 + 0 =$

(25) $13 + 10 =$

8) Add to 14, 15

(1) $14 + 1 =$

(2) $14 + 2 =$

(3) $14 + 4 =$

(4) $14 + 3 =$

(5) $14 + 5 =$

(6) $12 + 4 =$

(7) $14 + 6 =$

(8) $14 + 9 =$

(9) $14 + 7 =$

(10) $14 + 8 =$

(11) $14 + 9 =$

(12) $14 + 8 =$

(13) $14 + 10 =$

(14) $15 + 1 =$

(15) $15 + 2 =$

(16) $15 + 4 =$

(17) $15 + 5 =$

(18) $15 + 7 =$

(19) $15 + 3 =$

(20) $15 + 5 =$

(21) $15 + 8 =$

(22) $15 + 6 =$

(23) $15 + 9 =$

(24) $14 + 5 =$

(25) $15 + 7 =$

9) Add to 16, 17

(1) $16 + 1 =$

(2) $16 + 2 =$

(3) $16 + 4 =$

(4) $16 + 3 =$

(5) $16 + 5 =$

(6) $15 + 4 =$

(7) $16 + 6 =$

(8) $16 + 9 =$

(9) $16 + 7 =$

(10) $16 + 8 =$

(11) $14 + 9 =$

(12) $16 + 8 =$

(13) $16 + 10 =$

(14) $17 + 1 =$

(15) $17 + 2 =$

(16) $17 + 4 =$

(17) $17 + 5 =$

(18) $17 + 7 =$

(19) $17 + 3 =$

(20) $15 + 5 =$

(21) $17 + 8 =$

(22) $17 + 6 =$

(23) $17 + 9 =$

(24) $14 + 5 =$

(25) $17 + 10 =$

10) Add to 18, 19

(1) $18 + 1 =$

(2) $18 + 2 =$

(3) $18 + 4 =$

(4) $18 + 3 =$

(5) $18 + 5 =$

(6) $17 + 4 =$

(7) $18 + 6 =$

(8) $18 + 9 =$

(9) $18 + 7 =$

(10) $18 + 8 =$

(11) $15 + 9 =$

(12) $18 + 8 =$

(13) $18 + 10 =$

(14) $19 + 1 =$

(15) $19 + 2 =$

(16) $19 + 4 =$

(17) $19 + 5 =$

(18) $19 + 7 =$

(19) $19 + 3 =$

(20) $17 + 5 =$

(21) $19 + 8 =$

(22) $19 + 6 =$

(23) $19 + 9 =$

(24) $19 + 5 =$

(25) $19 + 10 =$

11) Add to 10-19

(1) $13 + 5 =$

(2) $15 + 2 =$

(3) $17 + 4 =$

(4) $19 + 3 =$

(5) $18 + 5 =$

(6) $12 + 4 =$

(7) $14 + 6 =$

(8) $15 + 9 =$

(9) $16 + 7 =$

(10) $18 + 8 =$

(11) $10 + 9 =$

(12) $13 + 8 =$

(13) $15 + 9 =$

(14) $11 + 6 =$

(15) $12 + 7 =$

(16) $19 + 2 =$

(17) $16 + 7 =$

(18) $19 + 4 =$

(19) $14 + 8 =$

(20) $15 + 5 =$

(21) $19 + 3 =$

(22) $18 + 6 =$

(23) $13 + 9 =$

(24) $17 + 7 =$

(25) $16 + 9 =$

12) Add to 10-19

(1) $14 + 5 =$

(2) $16 + 2 =$

(3) $18 + 4 =$

(4) $12 + 3 =$

(5) $11 + 5 =$

(6) $13 + 4 =$

(7) $15 + 6 =$

(8) $16 + 9 =$

(9) $18 + 7 =$

(10) $19 + 8 =$

(11) $11 + 9 =$

(12) $14 + 8 =$

(13) $15 + 9 =$

(14) $12 + 6 =$

(15) $10 + 7 =$

(16) $18 + 2 =$

(17) $17 + 7 =$

(18) $14 + 4 =$

(19) $15 + 8 =$

(20) $16 + 5 =$

(21) $18 + 3 =$

(22) $19 + 6 =$

(23) $11 + 9 =$

(24) $18 + 7 =$

(25) $17 + 9 =$

Timed Review 7!	(2 Points each)	Your Score:_____/100

10 + 5 =	10 + 8 =	10 + 9 =
11+ 9 =	17 + 1 =	16 + 6 =

11 + 3 =	11 + 4 =	18 + 5 =	17 + 6 =
12 + 8 =	15 + 7 =	18 + 9 =	18 + 6 =
12 + 5 =	12 + 3 =	18 + 2 =	19 + 3 =
12 + 4 =	16 + 5 =	12 + 6 =	12 + 7 =
12 + 8 =	15 + 9 =	12 + 7 =	18 + 5 =
13+ 2 =	13 + 4 =	13 + 2 =	16 + 3 =
13 + 4 =	17 + 5 =	13 + 6 =	13 + 7 =
13 + 6 =	18 + 3 =	13 + 9 =	17 + 5 =
14 + 5 =	19 + 1 =	16 + 2 =	19 + 3 =
14 + 4 =	14 + 5 =	15 + 2 =	10 + 0 =
14 + 8 =	14 + 7 =	14 + 3 =	10 + 9 =

19 + 1 =   14 + 2 =   18 + 3 =

18 + 2 =   14 + 7 =   19 + 2 =

13 + 3 =   17 + 4 =   14 + 5 =   13 + 6 =

12 + 6 =   10 + 7 =   14 + 4 =   17 + 5 =

15 + 7 =   12 + 6 =   16 + 2 =   14 + 5 =

14 + 4 =   15 + 4 =   16 + 6 =   19 + 5 =

12 + 4 =   12 + 5 =   17 + 1 =   19 + 2 =

13 + 3 =   13 + 5 =   17 + 3 =   19 + 4 =

11 + 4 =   17 + 5 =   17 + 6 =   19 + 7 =

11 + 8 =   16 + 6 =   17 + 8 =   14 + 5 =

15 + 1 =   14 + 4 =   18 + 2 =   15 + 9 =

15 + 4 =   15 + 2 =   19 + 3 =   13 + 0 =

14 + 3 =   15 + 3 =   18 + 2 =   11 + 8 =

$11 + 7 =$	$11 + 8 =$	$11 + 9 =$
$12 + 9 =$	$18 + 1 =$	$17 + 6 =$
$12 + 3 =$	$12 + 4 =$	$19 + 5 =$	$18 + 6 =$
$13 + 8 =$	$16 + 7 =$	$19 + 9 =$	$19 + 6 =$
$13 + 5 =$	$13 + 3 =$	$19 + 2 =$	$12 + 3 =$
$12 + 4 =$	$17 + 5 =$	$13 + 6 =$	$13 + 7 =$
$13 + 8 =$	$16 + 9 =$	$13 + 7 =$	$19 + 5 =$
$14 + 2 =$	$14 + 4 =$	$14 + 2 =$	$17 + 3 =$
$14 + 4 =$	$18 + 5 =$	$14 + 6 =$	$14 + 7 =$
$14 + 6 =$	$19 + 3 =$	$14 + 9 =$	$18 + 5 =$
$15 + 5 =$	$15 + 1 =$	$14 + 2 =$	$12 + 3 =$
$15 + 4 =$	$15 + 5 =$	$14 + 2 =$	$11 + 5 =$
$15 + 8 =$	$15 + 7 =$	$15 + 3 =$	$11 + 9 =$

$10 + 1 =$ $15 + 2 =$ $19 + 3 =$

$19 + 2 =$ $15 + 7 =$ $11 + 2 =$

$14 + 3 =$ $18 + 4 =$ $15 + 5 =$ $14 + 6 =$

$13 + 6 =$ $11 + 7 =$ $15 + 4 =$ $18 + 5 =$

$16 + 7 =$ $13 + 6 =$ $17 + 2 =$ $15 + 5 =$

$15 + 4 =$ $16 + 4 =$ $17 + 6 =$ $13 + 5 =$

$13 + 4 =$ $13 + 5 =$ $18 + 1 =$ $12 + 2 =$

$14 + 3 =$ $14 + 5 =$ $18 + 3 =$ $14 + 4 =$

$12 + 4 =$ $18 + 5 =$ $18 + 6 =$ $15 + 7 =$

$12 + 8 =$ $17 + 6 =$ $18 + 8 =$ $15 + 5 =$

$16 + 1 =$ $18 + 4 =$ $19 + 2 =$ $19 + 5 =$

$16 + 4 =$ $16 + 2 =$ $10 + 3 =$ $14 + 5 =$

$15 + 3 =$ $16 + 3 =$ $19 + 2 =$ $14 + 8 =$

13) Adding Tens

(1) $10 + 10 =$

(2) $10 + 20 =$

(3) $20 + 10 =$

(4) $10 + 30 =$

(5) $10 + 40 =$

(6) $20 + 20 =$

(7) $20 + 30 =$

(8) $10 + 50 =$

(9) $50 + 10 =$

(10) $10 + 60 =$

(11) $20 + 30 =$

(12) $30 + 30 =$

(13) $40 + 40 =$

(14) $50 + 50 =$

(15) $50 + 40 =$

(16) $30 + 40 =$

(17) $10 + 50 =$

(18) $60 + 40 =$

(19) $60 + 20 =$

(20) $60 + 30 =$

(21) $50 + 30 =$

(22) $50 + 20 =$

(23) $60 + 60 =$

(24) $20 + 50 =$

(25) $30 + 40 =$

14) Adding Tens

(1) $10 + 50 =$

(2) $10 + 60 =$

(3) $10 + 80 =$

(4) $10 + 90 =$

(5) $20 + 10 =$

(6) $20 + 40 =$

(7) $20 + 50 =$

(8) $20 + 60 =$

(9) $20 + 70 =$

(10) $20 + 80 =$

(11) $30 + 20 =$

(12) $30 + 30 =$

(13) $30 + 40 =$

(14) $30 + 50 =$

(15) $30 + 60 =$

(16) $30 + 70 =$

(17) $30 + 80 =$

(18) $30 + 90 =$

(19) $40 + 20 =$

(20) $40 + 30 =$

(21) $40 + 50 =$

(22) $40 + 60 =$

(23) $40 + 70 =$

(24) $40 + 80 =$

(25) $40 + 90 =$

15) Adding Tens

(1) $50 + 40 =$

(2) $50 + 50 =$

(3) $50 + 60 =$

(4) $50 + 70 =$

(5) $50 + 80 =$

(6) $50 + 90 =$

(7) $60 + 10 =$

(8) $60 + 20 =$

(9) $60 + 30 =$

(10) $60 + 40 =$

(11) $60 + 50 =$

(12) $60 + 60 =$

(13) $60 + 70 =$

(14) $60 + 80 =$

(15) $60 + 90 =$

(16) $30 + 70 =$

(17) $70 + 10 =$

(18) $70 + 20 =$

(19) $70 + 30 =$

(20) $70 + 40 =$

(21) $70 + 50 =$

(22) $70 + 60 =$

(23) $70 + 80 =$

(24) $70 + 90 =$

(25) $80 + 10 =$

16) Adding Tens

(1) $80 + 20 =$

(2) $80 + 30 =$

(3) $80 + 40 =$

(4) $40 + 80 =$

(5) $80 + 50 =$

(6) $80 + 70 =$

(7) $80 + 60 =$

(8) $80 + 80 =$

(9) $80 + 90 =$

(10) $90 + 10 =$

(11) $90 + 30 =$

(12) $90 + 20 =$

(13) $90 + 40 =$

(14) $90 + 60 =$

(15) $90 + 50 =$

(16) $90 + 70 =$

(17) $90 + 80 =$

(18) $90 + 90 =$

(19) $70 + 40 =$

(20) $80 + 40 =$

(21) $90 + 40 =$

(22) $30 + 50 =$

(23) $40 + 80 =$

(24) $50 + 90 =$

(25) $90 + 10 =$

17) Adding Tens

(1) $10 + 10 =$

(2) $20 + 10 =$

(3) $30 + 10 =$

(4) $40 + 10 =$

(5) $60 + 10 =$

(6) $50 + 10 =$

(7) $70 + 10 =$

(8) $90 + 10 =$

(9) $80 + 10 =$

(10) $90 + 10 =$

(11) $10 + 20 =$

(12) $20 + 20 =$

(13) $30 + 20 =$

(14) $50 + 20 =$

(15) $40 + 20 =$

(16) $60 + 20 =$

(17) $80 + 20 =$

(18) $70 + 20 =$

(19) $90 + 20 =$

(20) $10 + 30 =$

(21) $20 + 30 =$

(22) $30 + 30 =$

(23) $50 + 30 =$

(24) $40 + 30 =$

(25) $60 + 30 =$

18) Adding Tens

(1) $70 + 30 =$

(2) $90 + 30 =$

(3) $80 + 30 =$

(4) $10 + 40 =$

(5) $20 + 40 =$

(6) $30 + 40 =$

(7) $50 + 40 =$

(8) $40 + 40 =$

(9) $80 + 40 =$

(10) $70 + 40 =$

(11) $60 + 40 =$

(12) $90 + 40 =$

(13) $10 + 50 =$

(14) $20 + 50 =$

(15) $40 + 50 =$

(16) $60 + 50 =$

(17) $80 + 50 =$

(18) $30 + 50 =$

(19) $50 + 50 =$

(20) $90 + 50 =$

(21) $20 + 50 =$

(22) $30 + 60 =$

(23) $50 + 70 =$

(24) $40 + 80 =$

(25) $60 + 90 =$

19) Adding Tens

(1) $10 + 60 =$

(2) $20 + 60 =$

(3) $30 + 60 =$

(4) $40 + 60 =$

(5) $60 + 60 =$

(6) $50 + 60 =$

(7) $70 + 60 =$

(8) $90 + 60 =$

(9) $80 + 60 =$

(10) $90 + 60 =$

(11) $10 + 70 =$

(12) $20 + 70 =$

(13) $30 + 70 =$

(14) $50 + 70 =$

(15) $40 + 70 =$

(16) $60 + 70 =$

(17) $80 + 70 =$

(18) $70 + 70 =$

(19) $90 + 70 =$

(20) $10 + 70 =$

(21) $20 + 80 =$

(22) $30 + 80 =$

(23) $50 + 80 =$

(24) $40 + 80 =$

(25) $60 + 80 =$

20) Adding Tens

(1) $70 + 80 =$

(2) $90 + 80 =$

(3) $80 + 80 =$

(4) $10 + 90 =$

(5) $20 + 90 =$

(6) $30 + 90 =$

(7) $50 + 90 =$

(8) $40 + 90 =$

(9) $80 + 90 =$

(10) $70 + 90 =$

(11) $60 + 90 =$

(12) $90 + 90 =$

(13) $10 + 50 =$

(14) $30 + 10 =$

(15) $40 + 20 =$

(16) $60 + 30 =$

(17) $80 + 40 =$

(18) $30 + 50 =$

(19) $50 + 60 =$

(20) $90 + 70 =$

(21) $20 + 80 =$

(22) $30 + 90 =$

(23) $50 + 20 =$

(24) $40 + 50 =$

(25) $90 + 80 =$

(1) $10 + 40$

(2) $50 + 40$

(3) $10 + 70$

(4) $40 + 80$

(5) $90 + 80$

(6) $80 + 50$

(7) $70 + 40$

(8) $90 + 10$

(9) $20 + 80$

(10) $50 + 40$

(11) $80 + 40$

(12) $30 + 40$

(13) $30 + 90$

(14) $90 + 40$

(15) $80 + 40$

(16) $10 + 50$

(17) $60 + 20$

(18) $70 + 80$

(19) $70 + 50$

(20) $90 + 40$

(21) $80 + 10$

(22) $80 + 80$

(23) $50 + 70$

(24) $80 + 60$

(25) $70 + 70$

(1) 90 + 40

(2) 60 + 40

(3) 70 + 70

(4) 60 + 80

(5) 30 + 80

(6) 90 + 50

(7) 80 + 40

(8) 70 + 80

(9) 30 + 50

(10) 50 + 60

(11) 80 + 20

(12) 30 + 30

(13) 80 + 90

(14) 90 + 60

(15) 80 + 40

(16) 20 + 50

(17) 60 + 30

(18) 60 + 80

(19) 60 + 50

(20) 20 + 40

(21) 90 + 20

(22) 90 + 80

(23) 40 + 60

(24) 80 + 60

(25) 40 + 70

21) Add 2 digits by 1 Digit

(1)   16 + 3 =

(2)   25 + 3 =

(3)   31 + 8 =

(4)   18 + 1 =

(5)   16 + 1 =

(6)   55 + 4 =

(7)   53 + 5 =

(8)   72 + 2 =

(9)   25 + 4 =

(10) 33 + 5 =

(11) 17 + 2 =

(12) 26 + 2 =

(13) 40 + 7 =

(14) 80 + 3 =

(15) 61 + 6 =

(16) 53 + 5 =

(17) 40 + 6 =

(18) 32 + 7 =

(19) 12 + 2 =

(20) 16 + 3 =

(21) 23 + 5 =

(22) 91 + 3 =

(23) 23 + 5 =

(24) 64 + 3 =

(25) 71 + 5 =

22) Add 2 digits by 1 Digit

(1) $46 + 3 =$

(2) $45 + 3 =$

(3) $41 + 8 =$

(4) $54 + 3 =$

(5) $17 + 2 =$

(6) $51 + 4 =$

(7) $63 + 3 =$

(8) $82 + 5 =$

(9) $35 + 3 =$

(10) $33 + 6 =$

(11) $77 + 2 =$

(12) $96 + 2 =$

(13) $42 + 7 =$

(14) $81 + 3 =$

(15) $72 + 6 =$

(16) $54 + 5 =$

(17) $42 + 6 =$

(18) $32 + 7 =$

(19) $92 + 3 =$

(20) $26 + 3 =$

(21) $43 + 6 =$

(22) $81 + 8 =$

(23) $24 + 4 =$

(24) $65 + 4 =$

(25) $73 + 5 =$

23) Add Two Digits by 1 –Digit

You have mastered important skills.

Let's move on!

	(a)	(b)	(c)	(d)
(1)	2 8 + 0	3 7 + 1	4 6 + 2	5 5 + 2
(2)	6 4 + 3	7 3 + 2	8 2 + 3	
(3)	9 4 + 2	8 5 + 3	7 6 + 3	
(4)	7 8 + 1	9 1 + 1		
(5)	9 7 + 2			

24) Add Two Digits by 1 –Digit

	(a)	(b)	(c)	(d)
(1)	2 4 + 3	3 2 + 5	5 4 + 2	5 7 + 2
(2)	6 6 + 3	7 6 + 2	8 5 + 3	9 7 + 2
(3)	8 6 + 1	9 1 + 5	1 5 + 3	7 4 + 4
(4)	7 3 + 3	4 5 + 4	6 4 + 5	9 7 + 2
(5)	9 5 + 2	5 4 + 2	7 5 + 3	4 5 + 2

25) Add Two Digits by 1 –Digit

	(a)	(b)	(c)	(d)

(1)
```
  8 8        8 6        7 6        7 7
+   2      +   7      +   2      +   3
─────      ─────      ─────      ─────
  9 0        9 3
```

(2)
```
  6 6        6 7        8 7        9 6
+   3      +   4      +   5      +   5
─────      ─────      ─────      ─────
```

(3)
```
  9 9        8 7        8 7        8 9
+   2      +   6      +   8      +   2
─────      ─────      ─────      ─────
```

(4)
```
  5 5        6 7
+   9      +   3
─────      ─────
```

Great.
Keep Going!

(5)
```
  9 4        6 3
+   6      +   8
─────      ─────
```

26) Add Two Digits by 1 –Digit

	(a)	(b)	(c)	(d)
(1)	2 8 + 4	3 7 + 5	4 6 + 4	5 5 + 5
(2)	6 4 + 5	7 3 + 8	8 6 + 5	9 1 + 4
(3)	9 4 + 9	8 5 + 8	8 7 + 5	7 9 + 5
(4)	7 8 + 5	6 7 + 6	6 6 + 4	5 6 + 9
(5)	9 4 + 6	6 9 + 5	6 9 + 7	3 6 + 5

27) Add Two Digits by 1 –Digit

	(a)	*(b)*	*(c)*	*(d)*
(1)	8 8 + 4	8 6 + 5	7 6 + 4	7 7 + 5
(2)	6 6 + 5	6 7 + 4	8 7 + 5	9 6 + 4
(3)	9 9 + 4	8 5 + 4	8 7 + 5	8 9 + 5
(4)	5 5 + 5	6 7 + 5	6 5 + 4	6 9 + 4
(5)	9 4 + 5	6 3 + 5	9 8 + 4	2 2 + 5

28) Add Two Digits by 1 –Digit

	(a)	(b)	(c)	(d)
(1)	2 5 + 5	3 7 + 6	4 6 + 5	5 5 + 6
(2)	6 4 + 6	7 3 + 5	8 2 + 6	9 1 + 5
(3)	9 4 + 5	8 5 + 5	8 3 + 6	7 9 + 6
(4)	7 8 + 6	6 7 + 6	6 0 + 5	5 6 + 5
(5)	9 7 + 5	6 3 + 6	6 9 + 5	3 6 + 5

29) Add Two Digits by 1 –Digit

	(a)	(b)	(c)	(d)
(1)	8 8 + 5	8 6 + 6	7 6 + 5	7 7 + 6
(2)	6 6 + 6	6 7 + 5	8 7 + 6	9 6 + 5
(3)	9 9 + 5	8 5 + 5	8 7 + 6	8 9 + 6
(4)	5 5 + 6	6 7 + 6	6 5 + 6	6 9 + 5
(5)	9 4 + 6	6 3 + 6	9 8 + 5	3 2 + 6

30) Add Two Digits by 1 –Digit

	(a)	(b)	(c)	(d)
(1)	2 5 + 6	3 7 + 7	4 6 + 6	5 5 + 7
(2)	6 4 + 7	7 3 + 6	8 2 + 7	9 1 + 6
(3)	9 4 + 6	8 5 + 6	8 3 + 7	7 9 + 7
(4)	7 8 + 7	6 7 + 7	6 0 + 6	5 6 + 6
(5)	9 7 + 6	6 3 + 7	6 9 + 6	3 6 + 6

31) Add 3 Digits by 1 –Digit

Time to start new stuff!

	(a)	(b)	(c)	(d)
(1)	924 + 2	724 + 3	635 + 4	724 + 5
(2)	306 + 6	256 + 7	126 + 8	406 + 9
(3)	526 + 4	615 + 3	405 + 2	756 + 5
(4)	268 + 6	379 + 7	408 + 8	596 + 9

32) Add 3 Digits by 1 –Digit

	(a)	(b)	(c)	(d)
(1)	324 + 2	327 + 3	334 + 4	324 + 5
(2)	404 + 6	454 + 7	427 + 8	405 + 9
(3)	321 + 7	412 + 5	303 + 4	456 + 5
(4)	368 + 6	277 + 7	406 + 8	385 + 9
(5)	351 + 9	462 + 8	293 + 7	404 + 6

33) Add 3 Digits by 1 –Digit

	(a)	(b)	(c)	(d)
(1)	524 + 2	527 + 3	504 + 4	524 + 5
(2)	254 + 6	454 + 7	527 + 8	505 + 9
(3)	520 + 7	318 + 5	529 + 4	454 + 5
(4)	364 + 6	475 + 7	457 + 8	584 + 9
(5)	356 + 9	465 + 8	590 + 7	509 + 6

34) Add 3 Digits by 1 –Digit

	(a)	(b)	(c)	(d)
(1)	6 2 4 + 2	6 2 7 + 3	6 3 4 + 4	6 2 4 + 5
(2)	5 0 4 + 6	6 5 4 + 7	6 2 7 + 8	6 0 5 + 9
(3)	4 2 1 + 4	5 1 2 + 3		
(4)	5 6 8 + 6	6 7 7 + 7		
(5)	6 5 1 + 9	5 6 2 + 8		

Reading makes you smarter. Keep reading!

35) Add 3 Digits by 1 –Digit

	(a)	*(b)*	*(c)*	*(d)*
(1)	734 + 2	437 + 3	714 + 4	734 + 5
(2)	764 + 6	764 + 7	737 + 8	715 + 9
(3)	730 + 7	628 + 5	709 + 4	704 + 5
(4)	704 + 3	605 + 4	507 + 5	704 + 9
(5)	706 + 9	705 + 8	706 + 7	609 + 6

36) Add 4 Digits by 1 –Digit

	(a)	(b)	(c)	(d)
(1)	1 0 3 4 + 2	2 2 0 4 + 3	3 0 3 4 + 4	4 2 5 4 + 5
(2)	5 2 0 4 + 6	6 2 5 4 + 7	2 2 1 3 + 8	1 2 7 2 + 9
(3)	8 0 3 7 + 2	6 2 7 6 + 3	7 5 3 4 + 4	8 9 3 5 + 5
(4)	3 2 0 4 + 6	3 8 3 9 + 7	1 2 4 8 + 8	4 2 3 7 + 9
(5)	6 0 3 5 + 4	5 2 3 7 + 6	1 0 3 4 + 8	9 2 5 4 + 2

37) Add 4 Digits by 1 –Digit

	(a)	*(b)*	*(c)*	*(d)*
(1)	2035 + 2	3205 + 3	4635 + 4	5205 + 5
(2)	6205 + 6	7215 + 7	8235 + 8	9134 + 9
(3)	1038 + 2	2247 + 3	3035 + 4	4246 + 5
(4)	5205 + 6	6538 + 7	7239 + 8	8538 + 9
(5)	9036 + 4	1738 + 6	2035 + 8	3256 + 2

38) Add 2 Digits by 2–Digits

Get Ready! It is Time to Fly!

	(a)	*(b)*	*(c)*	*(d)*
(1)	2 0 + 1 3	3 0 + 2 4	4 0 + 3 5	5 0 + 4 2
(2)	6 0 + 2 4	7 0 + 2 5	8 0 + 1 6	9 0 + 1 6
(3)	9 0 + 2 7	8 0 + 2 7	8 0 + 1 9	7 0 + 3 9
(4)	7 0 + 2 8	6 0 + 4 5	5 0 + 5 3	5 6 + 6 3

39) Add 2 Digits by 2–Digits

	(a)	(b)	(c)	(d)
(1)	4 5 + 6 3	3 7 + 7 2	4 9 + 8 0	5 3 + 7 4
(2)	6 2 + 8 4	5 6 + 7 2	5 3 + 8 6	5 1 + 9 6
(3)	9 3 + 1 6	8 4 + 2 5	8 3 + 3 6	7 1 + 4 8
(4)	4 3 + 5 5	5 3 + 6 5	5 6 + 7 3	5 7 + 8 5
(5)	4 1 + 7 6	6 8 + 6 1	7 5 + 7 2	3 4 + 9 5

40) Add 2 Digits by 2-Digits at a High Speed!

	(a)	(b)	(c)	(d)
(1)	5 8 + 2 6	4 6 + 2 7	3 6 + 5 6	1 7 + 3 7
(2)	6 6 + 1 7	6 7 + 2 6	3 7 + 3 7	1 6 + 4 6
(3)	2 9 + 5 6	1 5 + 6 6	1 7 + 7 7	4 9 + 1 7
(4)	5 5 + 2 7	1 7 + 3 7	2 5 + 4 7	3 9 + 5 6
(5)	1 4 + 5 7	6 3 + 2 7	6 8 + 1 6	4 3 + 5 7

41) Add 2 Digits by 2–Digits

	(a)	*(b)*	*(c)*	*(d)*
(1)	71 +13	35 +24	46 +35	53 +52
(2)	62 +64	74 +75	85 +16	93 +16
(3)	96 +13	82 +27	85 +15	78 +12
(4)	73 +17	67 +13	50 +53	56 +60
(5)	47 +34	65 +17	69 +22	36 +36

42) Add 2 Digits by 2–Digits at high speed

	(a)	*(b)*	*(c)*	*(d)*
(1)	2 5 + 1 7	3 7 + 2 8	4 6 + 3 7	5 5 + 4 8
(2)	6 4 + 5 8	7 3 + 6 7	8 2 + 7 7	9 1 + 8 7
(3)	9 4 + 9 7	8 5 + 1 7	8 3 + 2 8	7 9 + 3 8
(4)	7 8 + 4 8	6 7 + 5 8	6 4 + 6 7	5 6 + 6 7
(5)	9 7 + 7 8	6 3 + 8 8	6 9 + 9 7	3 6 + 7 7

43) Add 2 Digits by 2–Digits

Time to speed up!

	(a)	*(b)*	*(c)*	*(d)*
(1)	2 8 +1 4	3 6 +2 5	2 6 + 3 4	3 7 +4 5
(2)	4 6 + 2 5	4 7 + 3 4	5 7 + 4 5	3 6 + 5 4
(3)	2 9 + 1 6	8 0 + 3 0	1 7 +2 5	
(4)	4 5 +6 5	3 7 +1 5	2 5 + 2 4	

44) Add 2 Digits by 2–Digits

	(a)	(b)	(c)	(d)
(1)	2 5 +1 5	3 7 +2 6	4 6 +2 5	5 5 +3 6
(2)	6 4 +4 0	7 3 +1 5	8 2 +3 0	9 1 +4 5
(3)	9 4 +4 0	7 5 +3 0		
(4)	7 8 +1 6	6 7 +2 6		
(5)	8 3 +5 0			

Keep Going!

45) Add 2 Digits by 2–Digits at higher speed

	(a)	(b)	(c)	(d)
(1)	7 5 + 1 6	4 7 +2 7	5 3 +2 8	6 7 +3 6
(2)	6 8 + 7 2	8 8 + 2 5	8 7 +3 9	9 6 + 8 5
(3)	9 5 +4 7	9 7 +3 6	6 5 +5 7	
(4)	7 3 +5 8	7 9 +2 6	6 9 +5 6	

46) Add 2 Digits by 2–Digits

	(a)	(b)	(c)	(d)
(1)	48 +1 1	5 6 + 1 2	6 6 +2 0	7 7 + 3 0
(2)	2 6 +4 0	3 7 +5 0	4 7 +6 0	5 6 +7 0
(3)	6 9 +1 1	8 5 +2 1	8 7 +3 1	3 9 +4 1
(4)	7 5 +5 1	8 7 +4 2		

47) Add 2 Digits by 2–Digits

	(a)	(b)	(c)	(d)
(1)	79 +60	60 +41	29 +50	75 +52
(2)	56 +35	69 +24	60 +24	98 +35
(3)	36 +65	36 +35	26 +25	
(4)	56 +27	69 +23	32 +46	

You are Getting Stronger?

48) Add 2 Digits by 2–Digits

	(a)	*(b)*	*(c)*	*(d)*
(1)	9 4 + 3 4	6 3 +1 5	9 4 + 2 5	6 3 + 2 3
(2)	2 8 +4 5	4 3 +5 5	2 8 + 6 4	9 6 + 2 6
(3)	9 7 +2 5	6 3 + 3 5	8 7 +1 7	
(4)	6 9 +4 5	6 7 +5 7	6 5 +8 7	

Great Work!

49) Add 2 Digits by 2-Digits

	(a)	*(b)*	*(c)*	*(d)*
(1)	2 5 + 1 7	3 7 + 2 3	4 6 + 3 4	5 5 + 4 5
(2)	6 4 + 5 6	7 3 + 6 7	8 2 + 8 7	9 1 + 7 0
(3)	9 4 + 9 1	8 5 + 2 3	8 3 + 1 6	7 9 + 6 8
(4)	7 8 + 4 0	6 7 + 2 8	6 4 + 4 3	

50) Add 2 Digits by 2–Digits

	(a)	*(b)*	*(c)*	*(d)*
(1)	8 4 +1 2	8 6 +2 5	7 6 +3 7	7 3 + 8
(2)	6 6 +5 0	6 7 +4 3	8 7 +2 8	9 6 +5 7
(3)	9 0 +1 7	8 5 +9 7	8 7 +4 9	8 9 +6 8
(4)	5 5 +5 6	6 7 +4 8	6 5 +3 8	

Practice is the key
to master math!

51) Add 2 Digits by 2–Digits

	(a)	*(b)*	*(c)*	*(d)*
(1)	2 5 +3 8	3 7 +2 5	4 6 +4 2	5 5 +3 2
(2)	6 4 +1 9	7 3 +2 8	8 2 +3 8	9 5 +5 0
(3)	9 4 +2 8	8 5 +6 0	8 3 +1 4	7 9 +6 2
(4)	7 8 +4 9	6 7 +6 2	6 4 +3 1	
(5)	9 1 +2 4	6 3 +3 5	6 9 +2 8	

52) Add 2 Digits by 2-Digits

	(a)	(b)	(c)	(d)
(1)	80 +25	86 +29	76 +48	77 +39
(2)	66 +49	67 +48	87 +39	96 +28
(3)	90 +48	85 +38	87 +29	89 +40
(4)	55 +19			
(5)	94 +30			

Let's Keep going! We are close to finish!

53) Add 2 Digits by 2-Digits

	(a)	(b)	(c)	(d)
(1)	8 8 + 1 7	8 6 + 2 8	7 6 + 3 7	7 7 + 4 8
(2)	6 6 + 5 8	6 7 + 6 7	8 7 + 7 8	9 6 + 8 7
(3)	9 9 + 1 7	8 5 + 2 7	8 7 + 3 8	8 9 + 4 8
(4)	5 5 + 5 8	6 7 + 6 8	6 5 + 7 8	6 9 + 8 7
(5)	9 4 + 9 8	6 3 + 1 8	9 8 + 2 7	4 3 + 9 8

54) Add 2 Digits by 2–Digits

	(a)	*(b)*	*(c)*	*(d)*
(1)	25 + 98	37 + 88	46 + 78	55 + 69
(2)	64 + 59	73 + 48	82 + 38	95 + 28
(3)	94 + 18	85 + 28	83 + 39	79 + 49
(4)	78 + 59	67 + 79	64 + 69	
(5)	97 + 29	63 + 39	69 + 18	

55)

	(a)	*(b)*	*(c)*	*(d)*
(1)	8 8 + 9 8	8 6 + 8 9	7 6 + 7 8	7 7 + 6 9
(2)	6 6 + 6 9	6 7 + 5 8	8 7 + 5 9	9 6 + 4 8
(3)	9 9 + 5 8	8 5 + 4 8	8 7 + 3 9	8 9 + 2 9
(4)	5 5 + 5 9	6 7 + 9 9	6 5 + 8 9	6 9 + 7 8
(5)	9 4 + 6 9	6 3 + 5 9	9 8 + 4 8	4 3 + 5 9

Wow! Excellent Work!

We are now starting higher Additions. Soon you will be done!

56) Add 3 Digits by 2-Digits

	(a)	(b)	(c)	(d)
(1)	124 + 52	127 + 53	104 + 64	224 + 85
(2)	204 + 86	154 + 37	127 + 38	205 + 59
(3)	220 + 67	118 + 75	129 + 84	154 + 15
(4)	264 + 26	175 + 37		
(5)	256 + 59			

57) Add 3 Digits by 2–Digits

	(a)	(b)	(c)	(d)
(1)	1 2 5 + 5 2	3 2 7 + 5 3	1 0 5 + 6 4	2 2 5 + 8 5
(2)	2 0 7 + 8 6	1 5 7 + 3 7	1 2 7 + 3 8	2 0 7 + 5 9
(3)	2 2 6 + 6 7	1 1 6 + 7 5	1 2 6 + 8 4	
(4)	2 6 6 + 2 6	1 7 6 + 3 7		
(5)	2 5 6 + 5 9			

58) Add 3 Digits by 2–Digits

	(a)	*(b)*	*(c)*	*(d)*
(1)	936 + 72	439 + 63	514 + 64	937 + 45
(2)	967 + 56	468 + 67	535 + 78	
(3)	908 + 97	407 + 15	506 + 24	
(4)	904 + 43	406 + 44		
(5)	736 + 82	745 + 98		

59) Add 3 Digits by 2-Digits

	(a)	(b)	(c)	(d)
(1)	1034 + 42	2204 + 53	3034 + 64	4254 + 25
(2)	5204 + 56	6254 + 47	2213 + 48	
(3)	8037 + 42	6276 + 53		
(4)	3204 + 76			

60) Add 3 Digits by 2–Digits

	(a)	(b)	(c)	(d)
(1)	2035 + 32	3205 + 43	4635 + 54	5205 + 65

(2)
```
  6205
+   56
```
```
  7215
+   67
```

Learning brings Success!

(3)
```
  1038
+   92
```
```
  2047
+   83
```

(4)
```
  5205
+   76
```
```
  6538
+   47
```

61) Add 3 Digits by 3-Digits

	(a)	*(b)*	*(c)*	*(d)*
(1)	465 +717	365 +426	465 +739	565 +848

(2)
```
   163          325          265
 + 867        + 889        + 896
```

(3)
```
   488          998
 + 730        + 752
```

(4)
```
   555
 + 696
```

Just Relax

(5)
```
   765
 + 620
```

62) Add 3 Digits by 3–Digits

	(a)	*(b)*	*(c)*	*(d)*

(1)
```
   8 2 3        8 3 3        7 2 7        9 6 5
 + 1 3 2      + 1 3 8      + 1 8 2      + 2 3 2
```

(2)
```
   5 6 7        8 2 5        7 6 5
 + 5 3 5      + 9 6 5      + 4 3 2
```

(3)
```
   4 8 9        8 6 9        3 8 6
 + 1 5 2      + 5 0 2      + 5 6 6
```

(4)
```
   3 9 8        8 5 7
 + 6 2 4      + 1 5 7
```

(5)
```
   8 9 3        5 8 3
 + 4 4 5      + 4 2 7
```

63) Add 3 Digits by 3–Digits

	(a)	(b)	(c)	(d)
(1)	665 +755	455 +639	897 +646	165 +686
(2)	565 +824	468 +644	485 +681	
(3)	604 +716	415 +627	935 +673	
(4)	845 +618	795 +618		
(5)	435 +648	976 +657		

64) Add 3 Digits by 3–Digits

	(a)	(b)	(c)	(d)
(1)	312 + 860	245 +257	175 +683	469 +719
(2)	152 + 967	524 +258	617 +686	549 +798
(3)		214 +256	176 +688	493 +709
(4)			Moving Faster!	617 +706
(5)				214 +546

65) Add 3 Digits by 3–Digits

	(a)	(b)	(c)	(d)
(1)	578 +977	578 +977	578 +977	578 +977
(2)	748 +973	748 +973	748 +973	748 +973
(3)	778 +974	778 +974	778 +974	
(4)	545 +886	463 +759	493 +489	

66) Add 3 Digits by 3–Digits

	(a)	(b)	(c)	(d)
(1)	324 +132	327 +523	334 +654	425 +235
(2)	404 +456	454 +650	407 +408	405 +507
(3)	320 +900	602 +685	703 +254	
(4)	903 +397	656 +876	749 +877	

67) Add 3 Digits by 3–Digits

	(a)	(b)	(c)	(d)
(1)	350 +461	527 +253	504 +324	625 +235
(2)	245 +142	415 +934	527 +915	505 +604
(3)	308 +608	318 +129	850 +465	
(4)	565 +427	843 +978	546 +870	

You have learned a lot
Already! Keep It Up!

68) Add 3 Digits by 3–Digits

	(a)	*(b)*	*(c)*	*(d)*
(1)	621 +420	622 +543	633 +654	625 +715
(2)	504 +826	654 +947	627 +138	605 +249
(3)	421 +364	512 +543	603 +652	
(4)	514 +786	549 +879	838 +797	

69) Add 3 Digits by 3–Digits

	(a)	(b)	(c)	(d)
(1)	2034 + 241	3204 + 352	4034 + 463	5254 + 524
(2)	6204 + 655	7054 + 746	8203 + 647	3272 + 428
(3)	9037 + 942	7276 + 853	8504 + 764	

Very close!

70) Add 3 Digits by 3–Digits

	(a)	(b)	(c)
(1)	4201 + 376	5002 + 257	
(2)	8206 + 356	9217 + 267	1236 + 548
(3)	3039 + 792	4041 + 683	5035 + 854
(4)	2135 + 439	7934 + 615	6243 + 926

71) Add 3 Digits by 3–Digits

	(a)	(b)	(c)	(d)
(1)	2 8	5 8	3 4	2 5
	1 0	5 1	2 3	3 5
	+ 1 2	+ 2 6	+ 1 2	+ 3 2
(2)	6 5	8 2	4 7	4 7
	4 3	1 6	2 5	2 2
	+ 2 2	+ 3 2	+ 4 0	+ 5 0
(3)	5 6	6 5	7 4	8 3
	3 3	4 4	5 5	6 6
	+ 6 2	+ 7 3	+ 8 4	+ 9 6
(4)	2 2	2 1	2 0	
	7 7	8 8	9 9	
	+ 3 3	+ 4 4	+ 6 6	

72) Add 3 Digits by 3–Digits

	(a)	(b)	(c)	(d)
(1)	5 2 8 2 1 0 +4 1 2	6 2 7 3 2 1 +5 1 3	7 3 8 4 3 2 +6 2 3	8 2 9 2 1 0 + 2 7 4
(2)	4 2 4 5 1 2 + 6 1 2	7 2 6 2 3 0 +4 1 2	4 2 8 2 4 1 +3 1 4	9 2 8 2 3 0 +3 1 3
(3)	1 2 3 4 5 6 +7 8 9	4 5 7 2 3 4 +5 6 7		
(4)	5 2 8 3 4 1 + 4 3 4	3 2 8 6 5 4 +5 6 0		

73) Post Test-: Add the Following.

1)
```
    6 2 7              7 9              2 0 7
  +     8            + 6 8            +   8 6
```

2)
```
    8 7 0 6            9 6              2 8 7
  +   3 5 6            7 5              0 4 3
                     +   2 3          +   2 6 7
```

3)
```
    2 8 7            1 7 0            3 2 8 0
    4 2 3            4 1 5            7 0 2 5
  + 6 5 8          +   2 6 8        + 7 2 6 8
```

4)
```
    2 9 8 3                           2 9 8 7
    8 0 2 3                           8 0 2 8
  +   3 6 8                         + 8 3 6 9
```

Congratulations!

For mastering all those Additions!

You have earned your Additions Certificate from your teacher!

Great Deal!

PROGRESS NOTES

PROGRESS NOTES

_____

_____

_____

_____

_____

_____

_____

_____

_____

_____

_____

_____

_____

_____

_____

_____

_____

_____

www.ingramcontent.com/pod-product-compliance
Lightning Source LLC
Chambersburg PA
CBHW081205180526
45170CB00006B/2219